HOW LIFE ON EARTH BEGAN

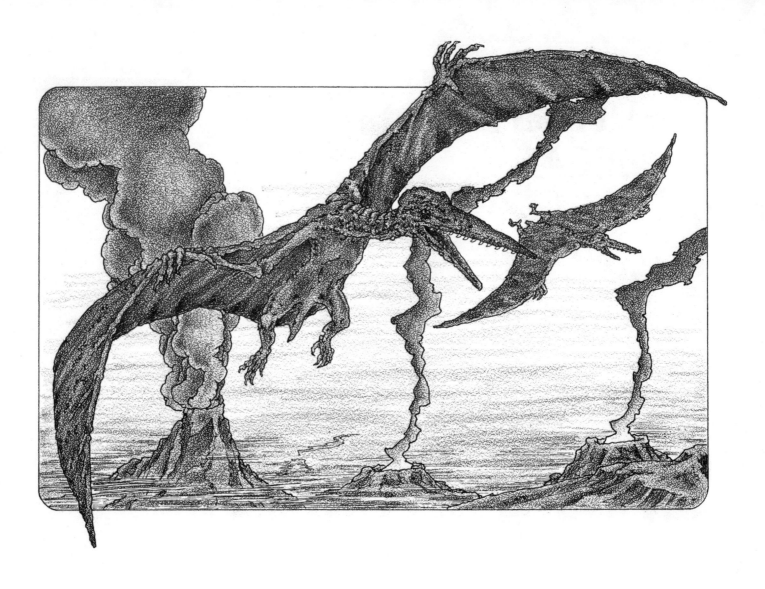

HOW LIFE ON EARTH BEGAN

by William Jaspersohn

illustrated by Anthony Accardo

Franklin Watts New York/London/Toronto/Sydney 1985

Library of Congress Cataloging in Publication Data

Jaspersohn, William.
How life on Earth began.

    Summary: Explains how life first appeared on the
earth as simple cells capable of reproducing them-
selves and then evolved into a series of more complex
organisms.
    1. Life—Origin—Juvenile literature. 2. Evolution—
Juvenile literature. [1. Life—Origin. 2. Evolution]
I. Accardo, Anthony, ill. II. Title
QH325.J33 1985   575   85-10557
ISBN 0-531-10030-8

For STEPHEN
with love and friendship

Has there always been life on Earth?
If not, where did we all come from?
And what was life on Earth like
before there was you and me?
To see how life on Earth began,
let us begin at the beginning.
Let us watch as the universe
is born, and then the planet Earth.

Scientists think that the universe
was born between ten and twenty
billion years ago.
They say it started from a *bang!*
Stars formed from the dust and gas
of this big bang explosion.

So did our Sun.
So did our planet, Earth.
The Earth and Sun were born
4.5 billion years ago.
At first, the Earth was a rocky planet—
cold, dry, and empty.

But slowly the Sun warmed the Earth.
And later, volcanoes erupted.
Steam and melted rock called *lava*
gushed through the Earth's crust.
The clouds of steam cooled
and turned to water.
The water fell on the young Earth.
The Earth was covered by a shallow ocean.
Other gases rose from the volcanoes, too.

Some, like nitrogen and hydrogen,
are in our air today.
Others, like ammonia and methane,
are not—they are poisonous.
But all these gases and others
lay over the young Earth.
They mixed together.
They made the Earth's first air,
or *atmosphere*.

As the Sun shone
on the young Earth,
the air and water
trapped the Sun's rays.
The Earth grew warmer.
Soon it was warm enough
for life to begin.
How did it happen?
How did life begin?
The ingredients for life
were in the young Earth's air.
Hydrogen, methane, ammonia,
and water—these are life's
simple building blocks.

When thunderstorms broke
over the young Earth,
their lightning bolts did
something amazing
to the gases.
The lightning changed the gases

into lumps called *molecules*.
Most of these molecules
fell with the rain into the oceans.
Time passed.
More molecules were made.
The oceans became a molecule soup.

By chance, some molecules
stuck together,
making chains
and molecule clumps.
While the molecules mixed together,
great quakes shook the Earth.
New lands rose from beneath the seas.
More time passed—
millions and millions of years.

And then there came a day
when, by chance,
a molecule was formed
that could make a copy of itself.
This copy-making skill is called
reproduction.
Molecules could now
reproduce themselves.
With nothing to stop them,
the reproducing molecules
filled the seas.
In time, some of them grouped
in a way
that formed a new kind of life
called a *cell*.
The first cells were born
about three billion years ago.

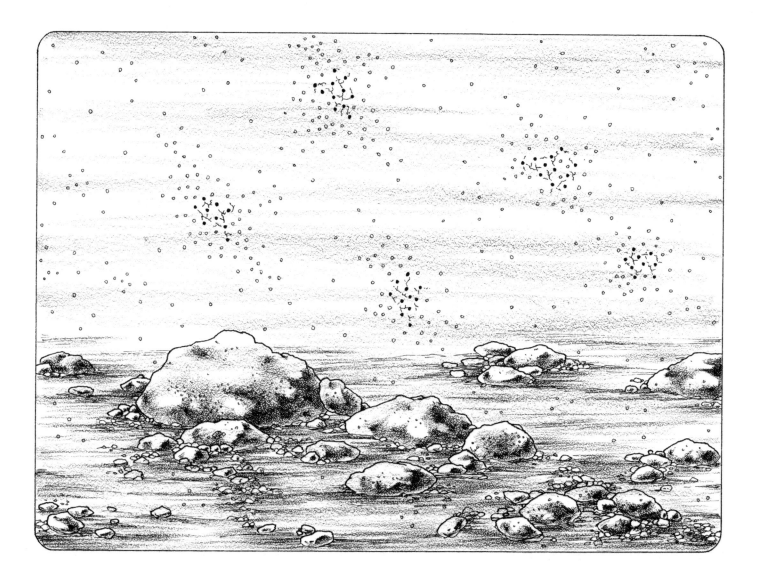

They were simple things,
smaller than dust specks.
The first ones couldn't even reproduce.
In fact, it took
two billion years for the first
reproducing cells to appear.
But when they did,
things started to happen.
New life came quickly
on the planet Earth.
The first reproducing cells
were simple *plant cells*.
They gave off a gas called *oxygen*
as they grew.
The oxygen mixed with the Earth's air.
Later, one-celled *animals*
breathed the oxygen.

Can you guess what happened next?
By chance, some of the plant cells lumped together.
So did some of the animal cells.
Millions of years and millions
of lumpings brought changes.

New kinds of many-celled plants arose.
And new kinds of many-celled animals
grew in the seas—
slender worms, rough sponges,
and squishy jellyfish.

Change from simpler to higher
forms of life is called
evolution.
Higher forms of life *evolve*
from lower ones.
As we have seen, life evolves by chance
over millions of years.
The animals that evolved
from the worms, sponges, and jellyfish
were the shelled animals.
Clams, corals, crablike creatures,
and sea stars—their shells
were something new
in evolution.
For the first time animals carried
their own shelter
from enemies.

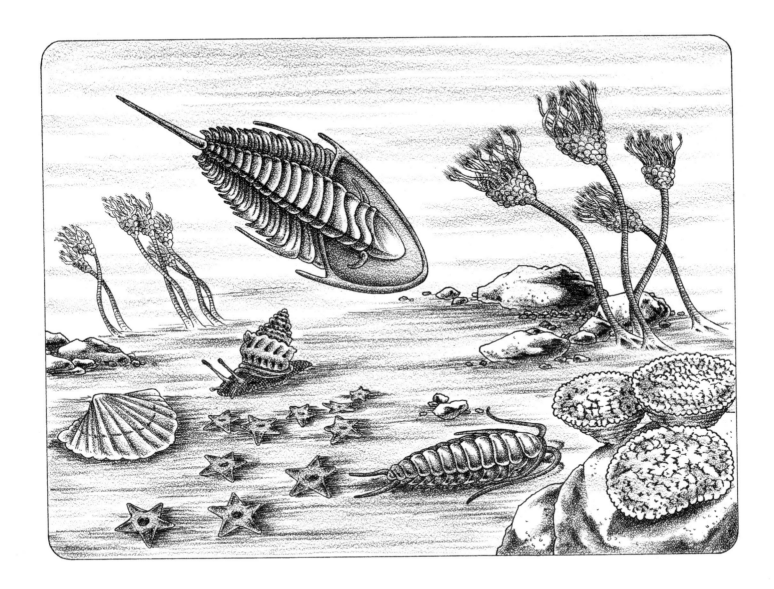

The first shelled animals appeared
about six hundred million years ago.
Soon after, the first fishes
appeared, too.
They had no shells.
Their skeletons were inside their bodies.
But they had speed.

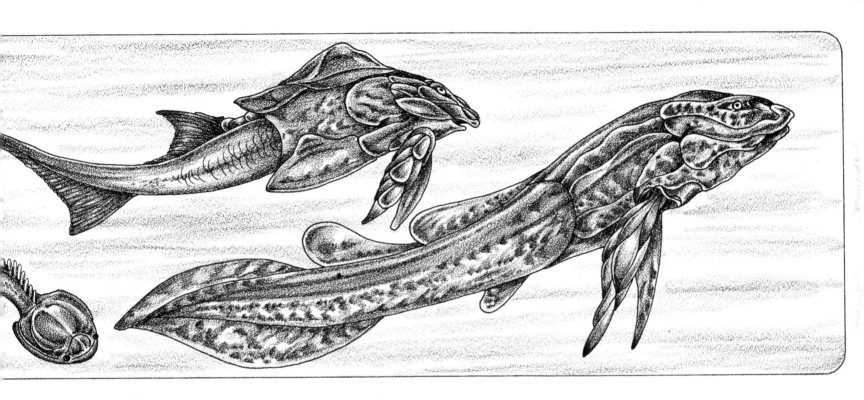

And they could swim wherever they wanted.
With these skills, the fish became
the rulers of the seas.
Time passed,
and more quakes shook the Earth.
More volcanoes erupted.
More new lands were formed.

Animals began exploring the land.
Fish were among them.
Some fish in those days
had lungs as well as gills.
They could breathe on the land.
They used their fins to crawl on the land.
Millions of these fish were born.
With each new fish,
the lungs became stronger.
The fins changed, too.
The fins became better
for walking on land.
After fifty million years or more,
the fins became legs.
The fish had evolved into
air-breathing animals that could walk.
Such animals as these
are called *amphibians*.

Life on land was hard
for the amphibians.
They died if their skins got too dry.
They still had to lay
their eggs in water.
New animals began to evolve
from the amphibians—

animals with thicker, stronger skins.
These animals did not have to live
in water.
They could lay their eggs on land.
They were *reptiles*.
They ruled the Earth
about two hundred million years ago.

The biggest reptiles were the dinosaurs.
They came in many strange shapes.
Some, like *Brontosaurus*,
were plant-eaters.
Others, like the ferocious
Tyrannosaurus rex, ate meat.

There were swimming reptiles
and flying reptiles.
One early flier was the
pterodactyl.
The birds we see today all evolved
from flying reptiles like it.

For millions of years
the Earth was a dinosaur paradise.
Then, seventy-five million years ago,
the dinosaurs disappeared.
No one knows what happened.
The weather might have changed—
grown hotter or cooler.
Plants might have died,
and with them,
the plant-eating dinosaurs.
Small animals might have died,
and without these for food,
the meat-eating dinosaurs died, too.
Most scientists think these changes
happened very, very slowly.
But finally, one day,
there were no more dinosaurs on Earth.

There were *mammals*.

They were different from dinosaurs.

They could live in different weather.

They cared for their young.

They had better teeth and jaws than dinosaurs.

And many of them were smarter.

They had stayed hidden
while the dinosaurs roamed the Earth.
Now these different mammals
began to evolve—into horses,
lions, bears, deer, dolphins,
and other mammals we see on Earth today.

One group that evolved
were the tree-dwelling mammals.
The earliest of these
looked like squirrels.
But time brought changes
to each new family of tree-dwellers.
Their paws became more like hands.
Their eyes became able to see colors.
Their eyes moved
to the front of their heads.
These changes happened slowly, as usual.
But after millions of years,
these new tree-dwelling mammals
could see and climb better.

The tree-dwelling mammals
evolved into monkeys.
This took thirty million years
or so to happen.
And twenty million years after that,
the monkeys evolved into apes.
The last in the line
of apes to evolve
was the ape-man
Australopithecus.
He was a short one,
less than five feet tall.
But his brain was big for his size.
That made him more intelligent
than his ape cousins.

And over a period of ten million years,
he evolved in many ways.
He moved from the forests to the plains.
He ran and hunted for his food.
He ate meat and used stones
for tools and weapons.
And yet, between one
and two million years ago,
Australopithecus vanished.
Why? Nobody knows for sure.
Maybe it was because
his brain stopped growing
and he couldn't evolve any further.
Nobody knows.
But a cousin of *Australopithecus*
did keep evolving.
He was *Homo*, which means *man*.
The earliest man was called
Homo erectus.

His larger brain made him even more
intelligent than *Australopithecus*.
Homo erectus made better tools and weapons.
He learned to use fire to cook food
and to keep warm.
He had no real fur,
so he made clothes for himself

from animal skins.
And he and his fellows talked.
They probably
made up words.
By talking, they
could pass on to others
what they knew.

Then an Ice Age came.
Great sheets of ice called *glaciers*
covered much of the Earth.
Now more than ever before,
Homo erectus survived
by using his brain.
He evolved.
With each new family,
the brain of *Homo* grew bigger.
Finally, one hundred thousand years ago,
the brain of *Homo erectus*
reached the size it is today.
A new, big-brained man had evolved.
He was thinking man,
or *Homo sapiens*.
Over the past hundred thousand years,
Homo sapiens has evolved
to the people we are today.

You are a *Homo sapiens*.
And so am I.
Our ancestors were the early apes.
And *their* ancestors were the early monkeys.
Before that, life stretches back
three billion years—
to an empty ocean, a thunderstorm,
some gas.
Life evolved by chance.
But it also evolved through struggle.
And the need to change and grow.
We are still evolving.
So long as we keep doing so,
our place on Earth remains.